Make: LOTS O' BOTS

by the makers of **Make:** magazine and Maker Faire

Scholastic Inc.

makezine.com/lotsobots

Series Editor: Michelle Hlubinka

Author: Marc de Vinck
Some text adapted from the writing of Kathy Ceceri, Andrew Terranova, Judy Castro, Ben Hylak, Mark Allen, Evil Mad Scientists Lenore Edman and Windell Oskay, Sylvia Todd, Craig Couden, Gareth Branwyn, Silvia Kacic, Zach and Kim DeBord, Robert Beatty, Dakota Peebler, Brad Peebler, Howard Wen, Bruce Stewart, Laura Cochrane, Gregory Hayes, Andrew Salomone, Phillip Torrone, Tod E. Kurt, Robert Luhn

All blue robot illustrations by Julie West.

Cover: Jeffrey Braverman (photography), James Burke and Juliann Brown (design)

Images by Maker Media, Make Labs, Douglas Adesso: 52; Theo Alexopoulos: 10; Artoo-Detoo.net: 22; Robert Beatty: 54; Jeffrey Braverman: 3, 12, 32, 43, 62; Judy Aimé Castro: 23; Ian Danforth: 27; Marc de Vinck: 56; Zach DeBord: 52, 53; Jerome Demers: 53; Tony DeRose: 6, 7, 8, 9, 62; Carla Diana: 21; Beth Doherty: 3; Kim Dow: 62; Kim Dow: contents, 62; Casey Duckering: 42; Lenore Edman: 14, 15, 16, 36; Evil Mad Scientist Laboratories: 14, 15, 16, 36; Don Feria for The Tech Museum of Innovation: 12; Tucker Gilman: 11; Dan Goldberg: 22; Vern Graner: 46; Arwen Griffith: (11,13); Mark Harrison: 35; Gregory Hayes: 19, 21, 46; Becca Henry: 28; Michelle Hlubinka: 4; Ben & Bridget Hylak: 24, 25; Bridget Hylak: 25, 26; Chris James: 22; Brian Jepson: 21; Zoran Kacic-Alesic: 44; Laewoo "Leo" Kang: 13; Kevin Kelly: 61; Erin Kennedy: 20; Gunther Kirsch: Make: builds; Timmy Kucynda: 53; Lindsay Lawlor: 40; Suzie Lee: 43; Jason Lentz: 12; Daniel Longmire: 46; Robert Luhn: 42; Mark Madeo: 38; Jeff Marinchak: (8,9); Sally Mason: 2; Garry McLeod: 21, 22, 33; Sabrina Merlo: 16; Chad Meserve: 21; Rob Nance: 43; Emily Nathan: 33; Andy Noyes: 32; Open Hand Project: 32; Windell Oskay: 14, 15, 16, 36; Brad Peebler: 37; Cody Pickens: 30; Robert Rausch: 34; Christian Ristow: 28; Beatty Robotics: 54, 55; Bruce Shapiro: 16; Della Shea: 11; Maker Shed: 53; Carla Sinclair: 49; Society for Science and the Public: 24; Jim St. Leger: 31; The Robot Group: 47; James Todd: 14, 15; Ugobe: 42; Riley Wilkinson: 4; WowWee Toys: 53.

No part of this publication may be reproduced, stored in a retrieval system, or transmitted in any form or by any means, electronic, mechanical, photocopying, recording, or otherwise, without written permission of the publisher. For information regarding permission, write to Scholastic Inc., Attention: Permissions Department, 557 Broadway, New York, NY 10012.

This edition is available for distribution only through the school market.

ISBN 978-0-545-74807-0

Copyright © 2014 by Maker Media, Inc. All rights reserved.
Published by Scholastic Inc., 557 Broadway, New York, NY 10012, by arrangement with Maker Media, Inc. SCHOLASTIC and associated logos are trademarks and/or registered trademarks of Scholastic Inc.
The Make logo and Maker Media logo are registered trademarks of Maker Media, Inc.

12 11 10 9 8 7 6 5 4 3 2 1   14 15 16 17 18/0

Printed in the U.S.A.   40

First printing, October 2014

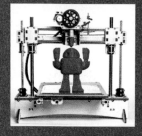

On the cover:
A **3-D Printer** is a kind of robot that will build, out of plastic, any design you can draw on the computer in a 3-D drawing program. In this picture, we're printing out our robotic mascot, Makey.

Contents

2   Let's Meet Lots o' Bots!

5   Everyone Is a Maker!

Chapter 1
6   Robots That Wow!
— Make: ScribbleBot

Chapter 2
18   Building Bot Buddies
— Make: BoxBot

Chapter 3
28   Robots Help Us
— Make: DusterBot

Chapter 4
38   Because We Can
— Make: VibroBot

Chapter 5
50   Join the Robot Revolution
— Make: Your Own Bot

60   Words Makers Know & Love

62   You Are a Maker! Join Us!

See this dot?

The **Make:** thunderbolt will zap you with lots more information! You'll find dozens of dots like this in this book. Go online to get more step-by-step instructions for any project marked with it.

makezine.com/lotsobots

Let's Meet Lots o' Bots!

In this book, you'll learn about all types of robots, or as we like to call them, "bots." We'll show you bots that are like pets and friends, and other bots that can make art or even help clean your room. Plus, we'll introduce you to some kids who make their own robots. Yes, kids can make robots too. In fact, we'll even show you how to make a few for yourself! So let's get started!

A Dalek is an archenemy of Dr. Who, the title character from the classic British science-fiction TV series. **Feuer Dalek** is a sculpture that moves around Maker Faire "exterminating all inferior lifeforms." These young visitors do look terrified!

Makers love robots for lots of reasons.

Robots help us play with ideas. Sometimes makers like the playful side of DIY (do-it-yourself) robotics: robots as toys or pets, robots that entertain. But bot builders are also getting serious, taking their robot-making from being a side hobby to their job.

The tools to make bots keep on getting easier to use and cheaper to buy. More people are making robots than ever before. You can be one of them! **Everything you need to make a robot is within your reach.**

A maker of a true robot needs to know about electronics, mechanics, and software. But we still call it a bot even if it's missing these things. Throw in a bunch of imagination and a bit of whimsy, and your robot is ready for Maker Faire!

What's Maker Faire? It's just the Greatest Show & Tell on Earth! At Maker Faire, makers show what they make, and share what they learn. It's a celebration of the Maker movement, and a great place to see lots o' bots, from little crocheted robots that you can hold in your hand, to a giant Dalek, to a Pancake Bot to pour batter into fancy shapes.

Whatever you make, bot or not, don't forget to tell us about it! We want to show off what you do at Maker Faire and in our magazines. What you make will inspire other kids to make too!

Beth Doherty's **Cro-bots** (above) may not have any electronics, but they have plenty of crocheted cuteness!

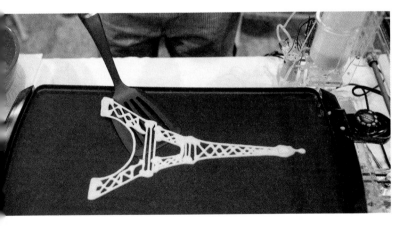

The **Pancake Bot** (above) has sizzled at Maker Faire since it first served off its hot griddle in 2010. In addition to circles, it can draw shapes and write words using pancake batter.

Be sure to check out our site for more about the bots in this book and even more information about making your own robot. Get step-by-step instructions for all the projects marked with the **Make:** thunderbolt.

Everyone Is a Maker!

Anyone can be a maker. You just need to have the right attitude! At Make, we like to say what the catbot in the cartoon on the opposite page says: "If you can imagine it, you can make it."

Every project in this book is something you can make. We know that's true because they're all made by makers who were once kids just like you—and some of them were made by kids!

Here's a test. Check to see if you are a maker. Look at the mosaic below. Using two pens, markers, or crayons (one color and the other black): **color** in the spaces that are true, and **black** out the ones that are false.

(If your colored spaces make an **M**, then you are probably a maker!)

1
Robots That WOW!

Sometimes makers build bots that are so cool and beautiful that they make us say "Ooo!" and "Aah!"

Robots can make us smile, they can make us laugh, and sometimes they even make us art. Many times the bot isn't even technically a bot, but that's OK! If it looks like a bot, smells like a bot, let's just call it a bot, even if a human is hiding inside working all of its levers, lights, and buttons.

Two teams of teenagers dreamed up, built, and programmed **Saphira**, an animatronic fire-breathing dragon. Read more about how they did it on the next page...

Saphira's steel-and-aluminum frame moves with air-powered, or pneumatic, controls. Saphira has five pistons: three for the head, one for the wings, and one for lifting the body of the dragon over a castle-like turret where she hides. A propane tank hooked up to the dragon's mouth lets Saphira breathe FIRE!

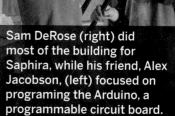

Sam DeRose (right) did most of the building for Saphira, while his friend, Alex Jacobson, (left) focused on programming the Arduino, a programmable circuit board.

Sketches for the original head and mechanisms

Saphira's new team updated the wings, added a robotic base, new eyes, sound, and more.

Some of Saphira 2.0's team

Named after a character in the book series Eragon, **Saphira** is a fire-breathing robotic dragon made by two teams of makers.

Two years after Saphira first appeared at Maker Faire, a second group of young makers tracked down the original team to ask how they built Saphira. They quickly adopted Saphira and started to bring her back to life.

Alex Jacobson, who programmed the original Saphira, has this advice to makers who are just getting started: "Just because other makers haven't done it before doesn't mean you can't. That is why I love making things and programming. If you can't find a program that will do what you want, just make your own!"

Maker Joseph DeRose uses a joystick and a laptop to control Saphira.

Saphira's second team built a larger castle wall for her.

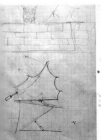
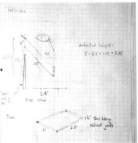
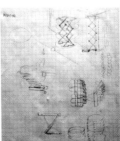

Saphira 2.0 wing designs

Jed Berk's **Blubber Bots** keep trying to escape, wandering out the door or up the back stairs. Or they drift like a sneaky group of cartoon clouds. What makes them move and react? Fans move the helium-filled Mylar balloon blobs. Light sensors send them toward the bright lights, while bump sensors tell them to turn around if they run into something. When Blubber Bots notice a cell phone, they whirl around and sing like whales. Wave your cell phone at one to say hello!

The brains of BlubberBots.

Kids investigate a Blubber Bot at the Beall Center in Irvine, California (right) and as part of a Build a Robot program at the Art Center for Kids in Pasadena (top.)

Pneubots are soft but strong; the giant-sized Ant-Roach model can easily transport several people on its back.

Robots can be soft, almost huggable. Saul Griffith and his company, Otherlab, make air-powered inflatables known as **Pneubots**. The "Pneu" sounds like "new" but comes from the word "pneumatic," which means they are air-powered. Inflatable robots move and change shape by adjusting the amount of air they contain, and they are strong enough to hold a basketball! This arm is kind of an air muscle that uses pneumatic control, just like Saphira.

Make a robotic balloon muscle! Lots of step-by-step projects on makezine.com/lotsobots

The air bladder inside a Pneubot was prototyped using rubber bicycle tubes. Pneubots look like giant beach toys. Their skin is similar to the material used to make bounce-house-type trampolines. Della Shea, the "chief seamster" for Otherlab, wrangles big bundles of plastic-coated nylon fabric into shapes like giraffes and elephants. She works with a team of engineers who are building a very slow, lumbering, air-powered robot. Pneubots can even box each other! (see right)

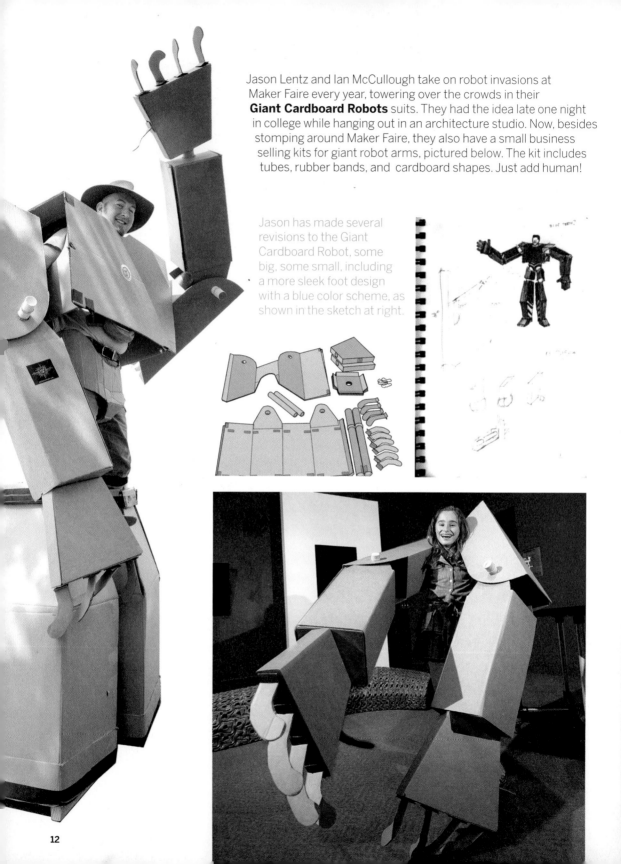

Jason Lentz and Ian McCullough take on robot invasions at Maker Faire every year, towering over the crowds in their **Giant Cardboard Robots** suits. They had the idea late one night in college while hanging out in an architecture studio. Now, besides stomping around Maker Faire, they also have a small business selling kits for giant robot arms, pictured below. The kit includes tubes, rubber bands, and cardboard shapes. Just add human!

Jason has made several revisions to the Giant Cardboard Robot, some big, some small, including a more sleek foot design with a blue color scheme, as shown in the sketch at right.

In artist Laewoo "Leo" Kang's sculpture **I Want To**, dozens of twitchy, little robots speak live Twitter **messages** out loud in their spooky, electronic voices. His program finds tweets that start with "I want...". Laewoo would like **us** think **about** why we want the things we want when we **watch what** his robots say and do.

Nitpickers may not call this nose-picker a robot, but why would we be so snotty?

Hop into a giant human hamster wheel to make it twist a finger into the nose of a 12-foot-high face. Tim Hunkin made a much smaller version of **The Disgusting Spectacle** as a coin-operated game decades ago. This version is bigger and grosser.

It's huge. It has nostrils. A finger goes in. Boogers come out. Eeeeew!

Maker SPOTLIGHT

Since age 8, **Sylvia Todd** has hosted her very own "Super Awesome Sylvia's Mini Maker Show" on how to make everything from crazy putty to a no-heat lava lamp.

A while ago, she came up with an idea for a robot that paints what you tell it to paint. She wanted to enter something surprising in a contest called RoboGames. She asked Windell Oskay and Lenore Edman of Evil Mad Scientist Labs to design it with her.

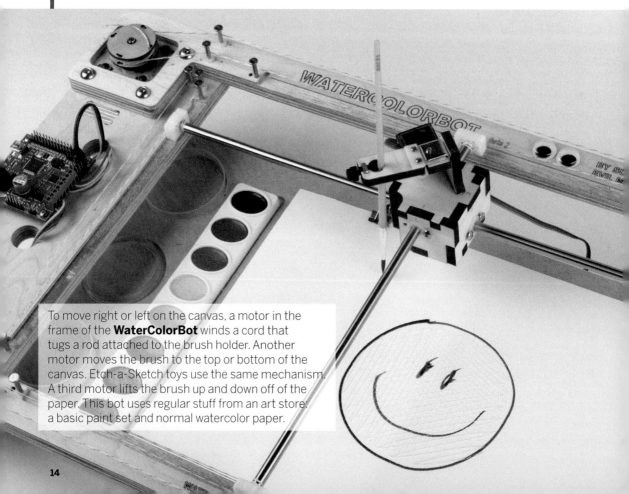

To move right or left on the canvas, a motor in the frame of the **WaterColorBot** winds a cord that tugs a rod attached to the brush holder. Another motor moves the brush to the top or bottom of the canvas. Etch-a-Sketch toys use the same mechanism. A third motor lifts the brush up and down off of the paper. This bot uses regular stuff from an art store: a basic paint set and normal watercolor paper.

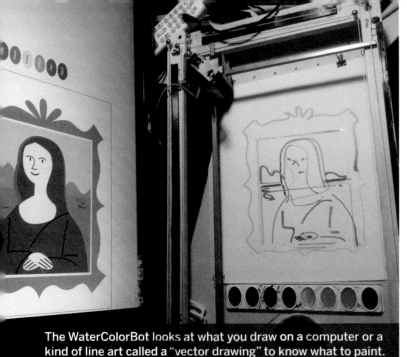

The WaterColorBot looks at what you draw on a computer or a kind of line art called a "vector drawing" to know what to paint.

Go online to watch "Super Awesome Sylvia's Mini Maker Show" to get lots of ideas for super-awesome things you can make at home.

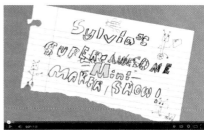

Once Sylvia, Lenore, and Windell started building the WaterColorBot, they realized that other people might like to have one as well, and so now they sell a kit that even you can make!

Sylvia travels the world sharing her WaterColorBot at Maker Faires, on TV, and even at the White House Science Fair! Here, she's at Maker Faire Kansas City.

The **Egg-Bot** does one thing very well: it draws on round objects. Use Egg-Bot to decorate eggs, lightbulbs, ornaments, balls, you name it! This patient painter can draw detailed designs easily. Bruce Shapiro made the original Egg-Bot kit with his friends Lenore Edman and Windell Oskay at Evil Mad Scientist Labs so anyone could make one of their own.

Make: SCRIBBLEBOT

Some bots wow us with how they look; others wow us with what they make. Like WaterColorBot, **ScribbleBot**'s only purpose in life is to make art with us. Unlike it, this bumbling bot moves in unusual ways, leaving random marks in beautiful patterns.

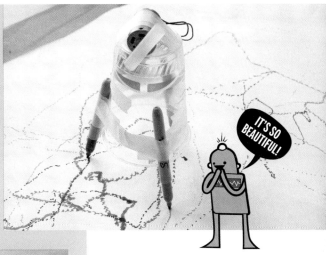

1 Add weight.

Add some weight to the motor shaft to offset it. You can try clay, wood, or a hot melt glue stick.

2 Attach motor.

Connect the motor to the battery. Attach the motor to the top of your base. Make sure your offset weight has the space to spin around.

3 Add pens.

Attach markers, pens, crayons, pastels, or even chalk to trace the jittering movement of your Scribble-Bot. (Optional: Use a steel wire to attach your pens to get more movement.)

Let it go!
Make some scribbles.

What You Need

PARTS

- Motor (1.5v-3v)
- Battery pack
- Sturdy base (small box, cup, cardboard, milk carton, berry basket)
- Something sticky and a little heavy (hot melt glue stick, clay, small binder clip)
- Large, flat canvas (mural paper, a big smooth piece of cardboard, chalkboard, dry-erase board)
- 3 things that draw (pens, chalk, markers, etc.)
- Masking tape
- Optional: steel wire

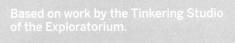

Based on work by the Tinkering Studio of the Exploratorium.

2
Building Bot Buddies

Makers build robots as a kind of "friend." Even the simplest of bots can move in unique or even repetitive ways that make us imagine that they just might have a mind of their own. What are they thinking? Can they really think? Robots will always tickle our imagination, and make us wonder.

Makers can bring stories to life by building replicas of our favorite robot characters that we've seen in movies or read about in books. They don't even need to be complicated to come to life. Just a bunch of recycled cereal boxes and a handful of imagination is enough to make your first buddy-bot.

Keepon may be the friendliest robot you'll ever meet. It's designed to be that way! Its simple face will make you smile, and its groovy way of squirming will make you want to dance along with it.

Doctors created Keepon to study how kids learn about making friends and growing up. Children who usually have a hard time talking to, looking at, or even being with other people practice their feelings and reactions with the original Keepon. Someone controls it from another room, and the cute robot views and records what the children do with a hidden camera.

Keepon's creators, Marek Michalowski and Hideki Kozima, sell a toy called My Keepon. It's less pricey because it had no remote control. Until now! Marek showed us a few simple steps to hack our My Keepons into animatronic puppets! Read how on the next page!

To control your My Keepon, all you have to do is connect the bot's guts to an Arduino. Then you can control it using just about anything you can imagine. Hacking your toys voids your warranty, but the fun that lies ahead is worth it! This sneak preview shows how easy it is to turn your toy into an animatronic puppet. Get full steps at makezine.com/lotsobots.

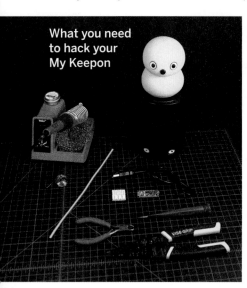
What you need to hack your My Keepon

1. Open the case.

2. Open the old brain.

3. Attach wires.

4. Widen the hole.

5. Close it up.

6. Connect the new brain.

True to her name, Erin "RobotGrrl" Kennedy of Montreal makes lots of different kinds of robots! She turns them into kits so that other robot-lovers can make them too. Her flock of **RoboBrrd**s frequently flutter into Maker Faire.

Spazzi bounces like Keepon, but it has a simpler way to move. Instead of four motors, it uses springs and solenoids. A solenoid is a special magnet that pulls a rod when you run electricity through it. This cute robotic bobblehead dances to music, and even makes some of its own. Learn how to make Spazzi from Marek Michalowski.

In the storybook *Leo the Maker Prince*, a girl meets a 3-D printing robot (above.) Along the way, the book shows how to use a 3-D printer to make many different things. Some things are fun and decorative. Others are practical and useful. You can even make the objects in the book on a 3-D printer if you want to!

The robots in this book aren't "real," but the author and illustrator of *Leo the Maker Prince*, Carla Diana, also works with real robots in the Georgia Tech Socially Intelligent Machines Lab. There, she designs the shells of robots, the outsides. The robots have eyes that blink, turn, and move around. They have heads that nod, turn, and tilt; and arms and hands that can hold, signal, and hug. Robot-makers call this kind of bot "humanoid" because it looks and moves a lot like people do.

In his series **Day of the Robot**, Chad Meserve makes scenes like those in the playful and quirky dioramas of Día de los Muertos. He puts robots in place of skeletons. He makes every figure, costume, and prop by hand. "I imagine all these robots stuck in the world we created, the world they took from us."

Dome R2 builders usually buy pre-made ones.

Sound Use the electronic guts of R2-D2 toys or a sound board with remotely triggered beeps!, blorps!, and tweets!

Material Build with plywood, plastic, fiberglass, resin, and aluminum.

Wheels To be able to control your R2 remotely, use 2 wheels per foot not just 1.

The Third Leg To add a mechanism that makes the extra leg go in and out, use a sturdy materials like lightweight metal.

The galaxy's most lovable robot inspires *Star Wars* fans worldwide to clone him. Meeting one of many **fan-made R2-D2s** feels like seeing a celebrity. That's because makers work from a common set of blueprints based on the actual R2-D2 movie prop. We cannot clone the Harrison Ford of 1977, but we can make thousands of Artoos to visit George Lucas fans around the world. That's pretty cosmic!

Making your own adorable droid takes hundreds of steps. Eight are shown above. First, R2 makers layer plastic, fiberglass, resin, and aluminum over a wooden frame.

After perfecting this display model, they add working lights and the character's familiar beep-boop sounds.

Next come radio controls for the wheeled legs and the turning, round head. The "2-3-2," moving R2's third leg in and out, poses the greatest challenge. The R2-D2 prop in the first *Star Wars* movie couldn't even pull it off.

A simple R2 can cost about $200 to build, but if you add features like a pop-up lightsaber chute, utility arms, a sensor scope, and a "holographic" projector, the tally can soar to $5,000 or more.

Besides attending science-fiction gatherings, R2s visit children's hospitals to cheer up the kids there. An R2 has to be on its best behavior. R2-D2 is a cute robot, not a killer cyborg, after all!

The funky little, free-range **Yellow Drum Machine** roams around, tapping away, recording its beats, and just having fun making music. It's not much bigger than your parent's shoe, but it's a whole lot smarter, and it smells better! A bunch of random parts he had collected inspired this drumbot's maker, Frits Lyneborg.

You don't need a bunch of shiny new materials to make an interesting robot. Judy Aimé Castro's **CoffeeBots** may look like they rolled right out of a junk heap, but she designed them to help her figure out what happens when you set loose a bunch of robots with a few small differences. It's a big idea scientists call chaotic behavior.

Judy's resourceful design makes a lot of robot with as little as possible: just a few parts, a little time and money, and a simple program. She teaches kids, so she wanted a bot that someone new to robotics and coding (writing the software that goes in the robots' brains) could make. Her students start with a simple code, and then mess around with the instructions to give the robots personality and style.

One CoffeeBot's program makes it either go to or away from light, depending on how it's wired up. Shine a flashlight, or hold your hand over a light sensor, and the bot will change what it does.

You can easily program a CoffeeBot to use its motors to move this way or that, or to blink its LED light in any pattern. Add sensors to measure distance or detect bumps or infrared LEDs. Your robots can also identify other robots and follow, lead, or talk with them!

Judy names each of her unique CoffeeBots after a different celebrity or historical figure. The way they act can create patterns resembling our own social behavior!

Born in Ecuador, Judy learned tinkering from her father, a machinist, and sewing from her mother. Today, she's an artist, industrial designer, and a teacher of makers-to-be.

 Make your own CoffeeBot with junk, some easy-to-find parts, and an Arduino! Step-by-step instructions for this and dozens of other projects on makezine.com/lotsobots

Maker SPOTLIGHT

Just weeks after he'd left the hospital after surviving months fighting a near-death illness, **Ben Hylak** was learning to solder at his first World Maker Faire in New York City. Still not strong enough to stand, he met dozens of makers from his wheelchair.

A year later, Ben (now fully healthy again) returned to Maker Faire as the maker of **MAYA**, his own telepresence robot. Telepresence robots let you be where you can't be in person, and MAYA stands for "Me And You Anywhere."

Ben says he hasn't done anything revolutionary, he just wanted to show people that a 13-year-old on a kid's budget could make a bot for 500 times less money.

Ben started ripping into old computers in his basement when he was 8. Then he'd put everything back together and make it work better than before. Just for fun!

As Ben's skills grow, his bots get more advanced. Ben's latest bot, **ALAIR** (Assisted Living Autonomous Internet Robot), takes the pulse, body temperature, and blood pressure of people living in nursing homes. ALAIR can even hand out the right pills to the grannies and grandpas in its care.

Having won lots of science fairs, Ben and MAYA were invited to meet President Obama at the White House Science Fair (above).

Part of Ben's design process is "user-testing": letting other people try out what you've made to figure out how to make it better (below and right).

Ben works on the robot's electrical panel. He cleverly used Lego pieces to hold the electronics in place (above.)

Ben works on his robots in a shed in his backyard (left). In this picture, he's testing ALAIR's patient tracking feature. The belt that Ben is wearing helps ALAIR detect its user.

Make: BOXBOT

Not all robots need to be made from electronics. They can be made of things you can find around the house, like cardboard and paper. Tiffany Threadgould showed us how to make her Cereal Box Bot during Maker Camp. Although it might not be a true robot, it sure will look like one, and it will be just as good of a buddy as any store-bought stuffed animal. Speaking of "looking" like a robot, you can make your **BoxBot** really come to life. How? You'll see, and so will your bot!

Lay out your boxes. Do you have a good mix of sizes and shapes? A tea box can be the head. The main part of the body can be a cereal or cracker box. The arms and legs can be long, skinny boxes, like a cracker box cut in half. Plastic cups or soda bottles can be arms and legs. Go collect more bot-body parts if you need them.

Create the main part of the robot body. You can hide the printed side of the box: just open it gently along all its glued seams, and then turn it inside out and reglue those edges you separated. If you want to add a radio or speakers or something else to your robot, add them now.

Punch a hole about a half inch down on the narrow side of the robot body near the top of the box. Then, punch a hole about a half inch down from the top edge of the tall skinny box.

Attach the arms to the robot body with a brad (paper fastener). Tape or glue the box lids closed. Repeat for the other side.

Attach the legs. Punch two holes in the bottom of the box, each about ¼ of the way in from the edge. Then punch one hole in each of two of the narrow boxes. Attach the legs with brads. Slide tape between the boxes to close all the lids.

What You Need

TOOLS

- Hole punch
- Scotch tape
- Scissors
- Ruler
- Markers, straws, and stickers for decoration
- Optional: small video camera or cameraphone with screen

PARTS

- Brads (paper fasteners)
- A big, empty box, like one that holds cereal, for the main part of the body
- A smaller, empty box, as from tea, for the head
- 2 long, empty boxes, as from crackers, for the arms
- 2 long, empty boxes or large cups for the legs
- Optional: zip ties (to attach "eyes" to BoxBot)

Attach the head. Punch a hole in the center of the top flap in the robot body. Then punch a hole in the center of the bottom flap on the robot head. Attach the head to the body with a paper fastener. Tape or glue the box closed.

Now, decorate! Punch holes in the head to add drinking straws for antennas. Put the finishing touches on your new BoxBot with stickers, markers, and anything else you can find!

Bonus add-on. Let's give this robot real eyes, like Fenn! If you have a camera phone with a big screen, record a short video of your eyes looking left, looking right, up and down, and making funny faces. Just film your eyes, not your mouth. Put the phone on the front of the head. Add some weight to the legs and body so your BoxBot won't tip over with the extra weight of the phone.

The idea for adding eyes to your BoxBot came from the **Fenn** (above.) This bot first showed up in a children's story about Earth getting in touch with another planet and the creatures who live there. Ian Danforth wrote the story and made the robot. He wants Fenn to be more like a pet than a machine.

3 Robots Help Us

Robots can help humans by going to places we don't care to go, or don't dare to go. They handle the "3 Ds" for us: dirty, dangerous, and dull. Robots can reach deep into the ocean or far into outer space. They go where it's too risky for humans to explore.

Robots can take care of a lot of practical things for us, too. Makers have built bots that knit us sweaters or tidy the yard. Robots don't need to sleep, rarely complain, and typically only eat a bit of electricity as they chug away, helping us explore new possibilities. Sometimes the bots we create can even start making other bots.

Christian Ristow's 7-ton, 26-foot-long **Hand of Man** picks up and crushes cars, pianos, motorcycles, and refrigerators. To move this giant's fingers, just use the glove (right.) You can lift about 3,000 pounds! Give a thumbs-up with the glove, and the giant robotic hand will give you a thumbs-up back!

Next time you need a rest while babysitting your neighbors or little siblings, why not break out Zvika Markfeld's **Bubblebot**? It blows big, wiggly soap bubbles that could make little kids (and you!) giggly for hours! Why should you be the one who has the dull job of dipping a wand and blowing? To make a giant bubble, people often tie cords into a loop between two sticks. A video of giant bubbles being blown on a Pacific beach in this way inspired Zvika to make the Bubblebot. On the Bubblebot, the sticks attach to a movable shelf that tilts up and down. After each dip, a special motor called a servo spreads the sticks and a fan blows air through the loop.

Maker SPOTLIGHT

"Don't be afraid to break something." This motto has served **Schuyler St. Leger** well for the past seven years as he's developed his making skills. When you *do* break something, it's time for superglue, or you need a robot that makes things. A **3-D printer** is a bot that does exactly that.

Schuyler first saw a 3-D printer when he was eight. A friend brought one to HeatSync Labs, his local hackerspace. "I was amazed!"

This electronic wonder builds real objects out of a thin line of melted plastic. Design anything three-dimensional and send the model to the printer. Like a good robot, the 3-D printer does what it's told and makes the object for you.

Once you create the design, the 3-D printer will make as many copies of the same object as you need. It's a lot like printing out lots of copies of a drawing on paper, but in three dimensions and in plastic.

But what can you do with a 3-D printing bot? The only limits are your imagination and 3-D modeling skills. You could replace that missing pawn from Mom's chess set, assemble the parts of a robotic hand, or even create robot-shaped chocolate molds. Recently, Schuyler used his bot to print a bunch of custom-made trophies for his robotics team.

Schuyler now owns two 3-D printers that create plastic objects. He hopes to get a bot that can make bigger objects with finer resolution. He dreams about the day when everyone has a 3-D printer that can sculpt objects in metal, carbon fiber, ceramic, and other materials, not just plastic. That day isn't too far off!

The Zen Gardener rakes sand into calming patterns you might see in a Japanese zen garden. It's something that humans usually do while thinking deeply about the mysteries of life. What do you think: Is it all right for a robot to do it for us instead?

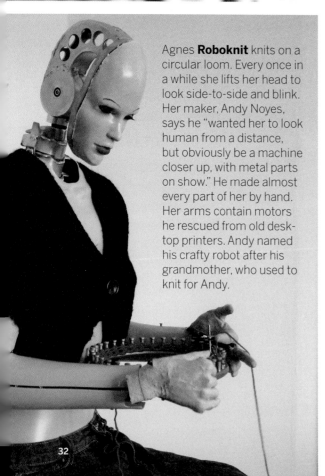

Agnes **Roboknit** knits on a circular loom. Every once in a while she lifts her head to look side-to-side and blink. Her maker, Andy Noyes, says he "wanted her to look human from a distance, but obviously be a machine closer up, with metal parts on show." He made almost every part of her by hand. Her arms contain motors he rescued from old desktop printers. Andy named his crafty robot after his grandmother, who used to knit for Andy.

A person who loses a hand in an accident will sometimes have it replaced with an artificial hand. Good ones can cost $100,000. In the **Open Hand Project**, hundreds of volunteers are working together to make a hand that can be made on a 3-D printer for just $1000, then used on people or on robots to do all the things a hand can do, like picking wildflowers.

Thanks to the over 10 million **Roombas** out there, vacuuming the old-fashioned way has become a thing of the past. But why let your Roomba just zip around and collect dust when you can use it to hack, modify, and take over the world?

Makers have figured out how to control them with game controllers and phones. A kid-owned company called **MyRoomBud** sells handmade costumes to turn them into frogs, tigers, and ladybugs. "If you don't dress up your Roomba, it's just a naked vacuum." Some makers make digital music from this vacuumbot. Others hack it to get snacks from the refrigerator. Just tired of letting it roam around willy-nilly? Hand over the controls to a hamster. Yes, a hamster! Rodents can run hacked Roombas from inside a plastic hamster ball with sensors.

In the **Feral Robotic Dogs** project, students of Natalie Jeremijenko take in abandoned robodogs and give them the special power to sniff out chemicals that hurt people and animals in places like this San Diego landfill.

Starting with the kind of toys that bark the national anthem or whiz in circles, they gently remove legs and put the dog bodies on used R/C cars with fat wheels for rugged off-roading. They add a new "nose": a chemical sensor. Each robot dog's brain tells it how to cover an area and what to do when it senses something toxic.

Activists all over the world, from Australia to Ireland, use these wild robotic pets to explore the local environment. Natalie's dogs are anything but toys.

J.D. Warren didn't like mowing his lawn. After he got hit by one too many rocks thrown from the back of his lawn mower, he started thinking, what if he could mow the grass from his back deck, or even the computer? **Lawnbot400** was born! His remote-control lawn mower uses an Arduino to navigate his acre of hills, dips, and rocks. What chore would you want a robot to do for you?

A **drone** can take many forms, but usually it is a flying robot. Most drones have electronics, power, and sensors in the middle, and arms with propellers pointing out of it. Using its propellers, a drone can lift into the air and fly this way and that.

Drones go by lots of other names, such as multicopters, multirotors, and UAVs (unmanned aerial vehicles). The number of arms gives each copter its name: a tricopter has 3 arms, and an octocopter has 8.

From the pyramids to the poles, makers all around the world use drones in clever ways. A 12-foot-long, solar-powered robotic boat called Scout got about a third of the way across the Atlantic (1,300 miles or 2,100 km) all by itself. Archaeologists send drones into skinny tunnels from the Queen's Chamber to other parts of the Great Pyramid in Giza, Egypt, to

make new discoveries. Park rangers and army soldiers in the mountains of Nepal use UAVs to help save wildlife. Some students in Sydney, Australia, may even get their textbooks delivered by air with a drone.

You can use your flying bot to take pictures from high up in the air, called aerial photography. Lighter drones, are safer and cheaper but they cannot carry as much stuff on them (like cameras and extra sensors), and their flying time is usually shorter.

The bigger your drone gets, the more scared you may be to fly it. Large drones can rip through clothes and skin, and they cost a lot of money, so be careful, and always have a parent, or at least a buddy with you! When you get started with drones, remember that the lighter the copter, the less damage it will do when it crashes. And it *will* crash!

The **Pool Noodle Drone** is a kind of unmanned aerial vehicle that you can fly using remote control, or you can set it off and let it use its own brain and sensors to buzz around the neighborhood. Its designer, Mark Harrison, used foam pool noodles so he could make the drone as cheap and as sturdy as he could. It laughs off any crash. And it floats, too!

Make: online

Make: DUSTERBOTS

Sometimes you need a little help, and a robot can be the perfect buddy to join in and save the day. A **DusterBot** will hunt down the dust bunnies under your bed. It'll buzz around helping you scrub your floors or desk.

You can't control where a DusterBot goes or what it does, but that's part of the fun. And forget about getting things done perfectly or quickly!

1. Cut handle.

Put on your safety goggles. Saw the handle off the hairbrush. You can ask an adult to help if you're unused to using a saw. Cut it close to the bristles, as shown here. Use sandpaper to smooth down the cut edge.

2. Attach motor.

Attach the motor to one end of the hairbrush using a cable zip tie. Cut off the extra length of zip tie.

3. Stick the pack.

Attach the battery pack to the brush using two-part hook-and-loop (Velcro) tape.

Lenore Edman and Windell Oskay of Evil Mad Scientist Labs got a lot of people excited about **bristlebots**. They showed millions of people how to use a vibrating motor and battery on top of a toothbrush. Researchers at MIT and Harvard use a similar bot design to study how birds form flocks or how fish swim in schools.

At a recent Maker Faire, Charley and Dakota Peebler and Stella and Giovanni Escobar-Rigon, all under 8 years old, taught visitors how to make all kinds of buzzy bristlebots that they called **HumBugs**.

What You Need

TOOLS

- Safety goggles
- Clamp
- Wire cutters/strippers
- Saw (hand or power—be careful!)
- Sandpaper or file

PARTS

- Hairbrush with soft bristles
- Hobby motor (you may find one in an old broken toy; ask before you take it apart!)
- Hook-and-loop (Velcro) tape
- Cable zip tie
- Optional: googly eyes, craft wire, pipe cleaners, and other crafty items to give your DusterBot some personality
- Optional: battery pack with batteries (or just a battery strong enough to make your motor move)

4 Connect leads.

Connect the battery to the motor. In this project, it doesn't matter which way you connect the wires (you can twist black to black, or black to red).

5 Set it loose!

Send your DusterBot off on its messy mission! Mod it for more fun! Add some personality: Connect two bots together with tape or a bit of wire. Trim or bend the bristles to make it turn.

Add wire legs to make it more stable.

4
Because We Can

Russell the Electric Giraffe stands about 17 feet tall with his neck raised. Turn the page to learn more about him and how he was made!

Many makers build robots to challenge their abilities, and push themselves to build bigger and better things. They enjoy learning new ways of making things, and luckily for us they also typically enjoy sharing what they have learned along the way. Don't be afraid to say "Yes, I can!" And if you have trouble, be sure to find other makers to help you build your idea. In fact, building bots with others is a great way to get started in the world of robotics.

When he first had the idea to make **Russell the Electric Giraffe** a decade ago, Lindsay Lawlor spent every moment of his free time for 10 solid months (and most of his money!) making Russell.

Russell the Electric Giraffe is never done: Lindsay's always changing the motors, software, electrical system, or the materials he uses. Russell is nearly alive.

Russell was named after Lindsay's engineer friend Russell Pinnington, who programmed Russell with 30,000 lines of code. The code gives Russell a brain, so he knows when people are touching his nose, and so he talks to people when they are near him.

Lindsay is a self-taught engineer. He never went to college or graduate school — he just reads the manuals and makes things work. For his giraffe, first he wrote the manual, then he built the beast. Just like Geppetto the puppetmaker from *Pinocchio*, Lindsay hopes that someday he can really bring Russell to life!

Russell's spots change colors and turn on and off to the beat of whatever music plays nearby.

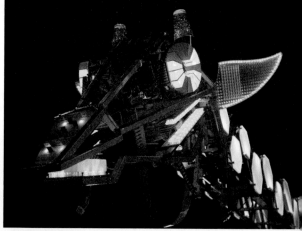

Makers are always looking to improve their bots, even if it means taking their head off and taking over the kitchen! Russell's creator, Lindsay, is never done working on Russell.

In 2014, Russell visited the White House for its first-ever Maker Faire

Russell's tires are more like shoes than wheels. He weighs about 2,000 pounds and can carry 9 children or 4 adults on his back.

Learn how to get this free pin featuring Russell the Electric Giraffe at makezine.com/lotsobots

Casey Duckering built this **Omni-Directional Robotic Hamster Ball** as a senior in high school. Tip a mobile phone to control where you'd like it to go. It can go in any direction at any time. A gyroscope, Arduino, and a 3-axis shifting drive mechanism control the ball's movements (right).

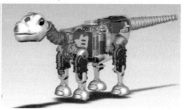

Ever-friendly **Pleo**, the robotic pet dino, is a love bot, designed to be your pal and to get you to cuddle him. Pleo is the size of a week-old baby *Camarasaurus*, a dinosaur from the late Jurassic period. Its naturally stocky legs hold perfectly all the motors and gears that make it move.

Caleb Chung (pictured left) and his team at Ugobe designed Pleo based on their research of actual dinosaur fossils. Pleo makes a great robo-buddy, but Chung also wants this little robotic dinosaur to answer a few big questions: Can a machine have a soul? Can it think, laugh and cry, bug you for a snack, tease you, or curl up and dream robotic dreams?

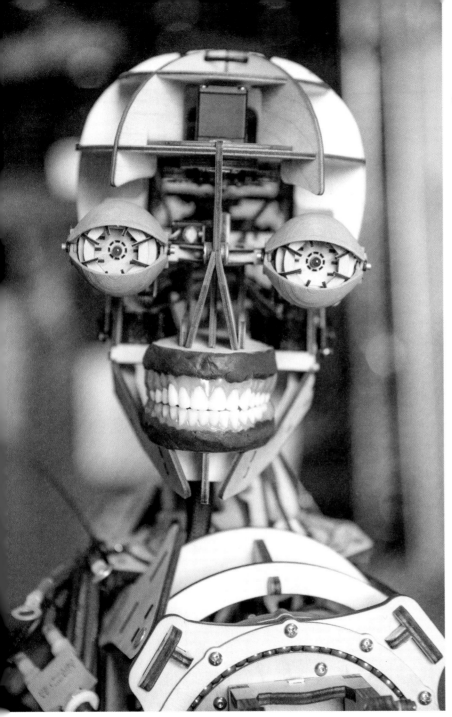

Sometimes a robot just doesn't want to do anything but take a nap. **The Useless Machine** has a long history. It was originally invented in 1952 by Marvin Minsky, a researcher in artificial intelligence. Once you turn it on, the machine quickly turns itself off. That's it! It's a truly useless machine.

Robots aren't always made of metal or plastic! **Roy** (above) is made of plywood cut on a lasercutter.

The **Walking Pod** (left) uses legs inspired by the Strandbeests by Theo Jansen, just like Russell the Electric Giraffe. Inside is a safe place to live and work: a little mobile home—that can take a stroll!

Maker SPOTLIGHT

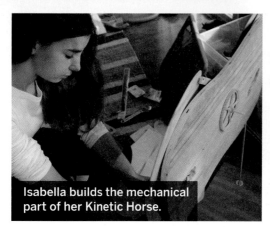

Isabella builds the mechanical part of her Kinetic Horse.

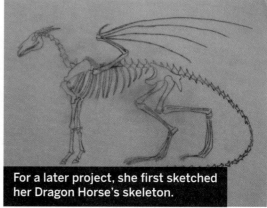

For a later project, she first sketched her Dragon Horse's skeleton.

Isabella Kacic would like to make a full-size horse robot someday. She's been riding for years, and she loves art. She also likes taking apart her robotic toys. At 13, she brought all those hobbies together by building a **Kinetic Horse**. "Kinetic" means that it moves instead of just sitting there.

To make it, Isabella studied where a horse's muscles and limbs go when it walks, trots, canters, and gallops. It was a challenging project. "The grown-ups told me that this was very difficult to do and had never been done, so I went on the Internet and found these cool kinetic sculptures by Theo Jansen and decided that doing something like that would be a good start."

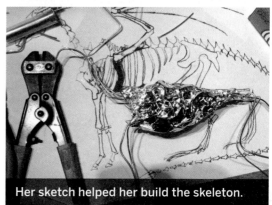

Her sketch helped her build the skeleton.

Isabella keeps on making. For her latest Maker Faire project, she returned to her theme of horses and mechanics, adding an Arduino and a sensor to activate the flapping wings of her **Dragon Horse** sculpture. Neither one of these is big enough to ride, but Isabella's on her way to her goal of a robot horse.

Next, she added plaster to the skeleton.

Isabella's first model of her Kinetic Horse (lower left) used K'NEX pieces. She then tested a design that used art board. Her beautiful final wooden version, in the back, looks like a Paint Mustang, a breed she rides during equestrian lessons.

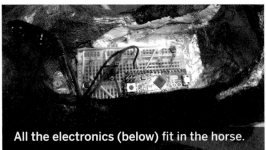

All the electronics (below) fit in the horse.

Isabella's final Dragon Horse looks great!

As a teenager, Lisa Winter rode to fame with her own fighting robot, **Tentoumushi** (pictured above with Lisa and her dad, Mike). Its cheerful ladybug shell hid a deadly "smothering dome" that defeated most of its opponents until Lisa and her robot ranked sixth in the world in *Robot Wars*, the late 90s show. Now at age 27, she builds robots as an inventor and artist at the company Stupid Fun Club, below.

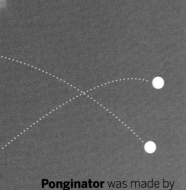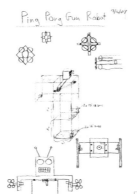

Early sketches of Ponginator

Ponginator's head is just a big, blue storage bin turned upside down.

An earlier version of Ponginator

Ponginator was made by a group of bot-making friends in Austin, Texas, called The Robot Group. Vern Graner had the idea, and everyone else asked, "How hard could it be?" That's exactly the kind of question that will get you into loads of trouble, and lots of fun, too.

Ponginator does two things: he scares people by launching Ping-Pong balls and by being big. From a height of 20 feet (7 meters), Ponginator showers balls down on passersby. But don't be afraid! Those balls shoot out very fast, but they wouldn't hurt a fly. The moment the balls leave Ponginator's arms, they slow down quickly. In this picture, Ponginator was sent up a crane lift to be even more terrifyingly silly.

Make: VIBROBOT

Inspired by the wonderful windup toys by the Brazilian artist Chico Bicalho, Mark Frauenfelder made this **VibroBot** with his daughter. You can make one in an hour or two. It's got the long legs of a giraffe. Add a lightweight neck, maybe a little LED bling, and you've got a mini Russell the Electric Giraffe!

❶ Punch 3 holes.

Use the hammer and screwdriver to punch a leg hole through one end of the bottom of the tin. Punch another hole on the other end of the bottom of the tin. You'll use these holes to attach the legs. Punch a hole through the lid near one end. This hole is for your wires.

❷ Make the legs.

Snip off two long pieces of wire from a coat hanger. Bend each one into a V-shape. Bend the tip of the V into a right angle. Bend a little "foot" at each end, and add a drop of hot glue to each foot to give them rubber tips.

❸ Attach the legs & weight.

Attach the legs to the holes in the tin using bolts, nuts, and metal washers. To add an offset weight, push a paper clip through one of the plastic flat washers, and attach the washer to the spindle of the motor.

❹ Install the motor.

Put two plastic flat washers on the lid under the motor. Put the battery pack under the tin. Tighten the motor and battery pack onto the tin using a cable tie.

What You Need

TOOLS
- Hammer
- Hot glue gun
- Phillips screwdriver
- Soldering iron (optional)
- Wire cutter/stripper

PARTS
- Metal candy tin
- Wire coat hanger
- Paper clip
- Plastic washers (3)
- Cable zip tie
- Batteries
- Wire, about 1 foot
- Nuts and bolts (small)
- Metal washers (small)
- Motor, 1.5V (can be recycled from a battery-powered toy)

5. Set it loose!

Attach the battery pack wires to the leads of the motor. (You may need to solder them.) Gently bend the paper clip and legs into different shapes and see how your robo-critter changes what it does.

5
Join the Robot Revolution

The thought of designing and building your first robot could seem a bit scary. Well, it's not!

The best way to get into robotics is to jump right in and get making. Your first robot might not help you fold your clothes or clean your room, but it just might be able to buzz around the house and entertain your cat. The important part of your robot-building journey is to just get started. This chapter will help you do just that!

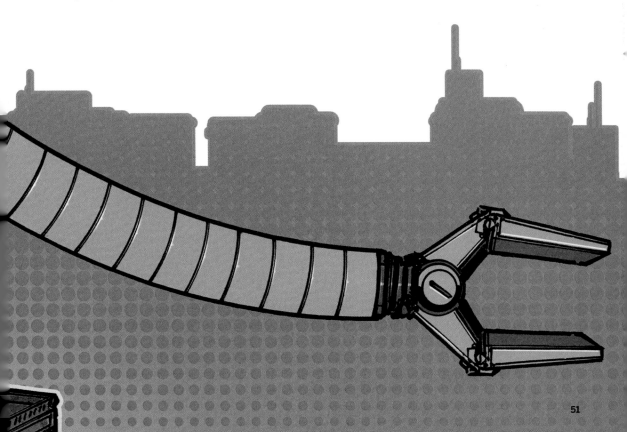

How simple can a robot be and still be a true robot? Back in the 1970s, insects and other critters with teeny brains inspired MIT researcher Rodney Brooks to design bug-like bots. Rodney's bots used very little computing power to do just what much more complicated robots (which we'll call "fussybots") could do.

A robot-maker named Mark Tilden took Rodney's idea even further. Mark wondered if he could create simple robots that needed no brains at all. The answer was yes, and **BEAM** robotics was born.

The letters of BEAM stand for biology, electronics, aesthetics, mechanics. (Aesthetics means making things look nice.) With BEAM, you make a bot that can see the world around itself and react in some way, with no need for any programming.

BEAMers (people who love BEAM robots) hack common household electronics in new, surprising, and more efficient ways. BEAM bots compete against one another in races, sumo matches, high jumps, rope climbs, and other Olympics-style events.

Over the past two decades, BEAMers worldwide have come up with lots of cheap and easy-to-build bots, like rollers, spinners, walkers, fliers, squirmers, and sitters. They all have just a few simple electronic parts, clever circuits, and bits of techno-junk to create cool robo-critters—many of them inspired by nature.

BEAM technology uses parts hacked from common household objects. BEAMers say, "My many dumb bots beat your one smart bot." They imagine a future with swarms of cheap robots cleaning skyscrapers, toxic waste sites, and even your house. If a few of them break or run out of energy, that's fine. When a single high-priced fussybot breaks even a little bit, you're out of luck.

BEAM us up, botties!

 Real insects have simple brains, and so a lot of BEAM bots act like insects. You can make a **beetlebot** that can avoid obstacles on the floor without any programming! Two motors move the bugbot forward. When one of its feelers hits something, the bot turns around. All you need to build the beetlebot is two switches, two motors, and a battery holder!

The **Herbie the Mousebot** kit (below) is an easy-to-make, light-chasing robot. These little bots are so fast, you have to run to keep up with them. We found that out the hard way when one ran away from us!

Robosapien (above) is an impressive BEAM bot. It's packed with technology that allows it to walk, run, pick up and throw things, and a ton of other interesting functions. It can even perform a little kung-fu when the time comes to protect your lab.

Mousey the Junkbot is a BEAM robot you can make out of an old computer mouse. This little "Frankenmouse" scoots quickly across the floor, thanks to lively little motors. And when the critter crashes into anything, it speeds off in the opposite direction.

Maker SPOTLIGHT

Genevieve and Camille were just 9 and 11 when they convinced their dad, **Robert Beatty**, to start building robots together at their home in western North Carolina. Genevieve was inspired by the "droids" she saw on TV. Camille was and still is inspired by nature. A documentary about NASA's rovers *Spirit* and *Opportunity* made them want to build their own. For the Beattys, building robots is all about the inspiration.

None of them had ever made robots or other electronics, so the Beattys learned everything they needed online as questions came up. Both girls are involved in design, soldering, electronics, metal fabrication, programming, naming, and testing of the robots. Dad guides the way. They even make their own tools, like a homemade computer-controlled cutting machine, a "CNC," to create their own metal parts! "The vertical mill is good for modifying parts," Camille says. "If we need to make a part from scratch, we use the CNC."

Lunokhod (right) means "Moon Walker" in Russian. Beatty Robotics built a functional one for a space museum in Prague.

Genevieve, also known as "Julajay," does 95 percent of the soldering on the family's projects. She also likes using their CNC.

One of Camille's favorite projects is **Snailbot** (above), a seashell with the robotic innards stuffed inside the shell. Camille is inspired by the idea of combining nature with technology. Also known as "Lunamoth," Camille enjoys doing most of the assembly work. She machines parts on their vertical mill, and designs and cuts parts on the CNC. Camille also enjoys making miniature robots as small as she can make them.

The girls and their dad write a tech blog about their robotics adventures. (Read about them at beatty-robotics.com!) The New York Hall of Science came across it, and they asked them to build a functional Mars Rover (above) for a permanent exhibit where kids can drive the robot around by remote control and search the Mars-scape for signs of life. The girls also built a second rover. They take it to children's hospitals so that sick kids who can't leave the hospital might get to drive the rover themselves.

Aluminalis, a 16-legged crawler, can roam on her own around the workshop, using sound sensors to find the best path and avoid bumping into anything. She responds to requests but prefers running on her own. She can be very difficult to catch! When in "shy mode," she moves away as you get closer, always finding a spot halfway between you and the rest of the room. Aluminalis consists of 846 individual components, many machined by the Beattys.

Make: YOUR OWN BOT

Play with your ideas, sketch out some concepts, and grab some junk from the kitchen drawer or parts from an old toy and see what you can make. Be sure to pay attention to what works and doesn't work. A robot-building journal is a great place to write down your thoughts and keep track of your creations.

Where to Start? It can be difficult to think about where to start when designing your own robot. Sometimes it's best to just sit down and start imagining. Make a list. Start with the things you want your bot to do. Will it be your friend? Does it need to talk, throw a ball, or ride a bike? If you can imagine it, you can make it.

Just remember to follow these 3 simple rules by Isaac Asimov when designing your bot:

1. A robot may not injure a human being or, through inaction, allow a human being to come to harm.

2. A robot must obey the orders given to it by human beings, except where such orders would conflict with the First Law.

3. A robot must protect its own existence as long as such protection does not conflict with the First or Second Law.

OK, most robots wouldn't break any of these rules, but you never know what the future might bring, right?

IDEA/PROJECT		DATE	
NOTES/SIG		FROM PAGE	TO PAGE

Get Sketching! Sketching is a perfect first prototype, and it only requires a pen or pencil and some paper to get started. Why sketch? Well, if you are working with a friend, now you both know what it should look like. If you are working alone, it's always good to have something to look at in case you get stuck. Remember, your bot can look like anything. It might not have a head, or it might only have a head?! Think about how it will move, how it will work, talk, and play.

Sketch an idea here!

As my cyborg friend would say:

"So what are you waiting for? The robots aren't going to build themselves. Yet."

— Gareth Branwyn

DATE

FROM PAGE

Make: Cool Stuff

And another one here!

Start Making! Once you've defined your bot and made some sketches, it's time to start construction. Collect some parts, gather some friends, and get building. The Internet is a great place to learn all about how your robot can move, think, talk, climb, or run. Check out makezine.com and search for "robot." You'll be amazed at how many bots are listed in the results. Even more amazing is how many have instructions on how to build it yourself.

Don't stop now! Keep sketching!

Maker:

Words Makers Know & Love

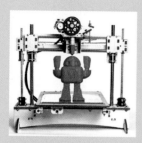

3-D Printing
Building real objects out of a thin line of melted plastic. 3DP is short for "3-D Printing."

Arduino
A kind of microcontroller.

Autonomous
Able to do things without anyone's help. An autonomous robot doesn't need someone using a remote control to tell it what to do. It has its brains on board.

CNC
Short for "computer numerical control," a type of machine or tool whose movements are precisely controlled through computer code.

Code
The list of steps you want a robot (or computer) to take in the order you want it to take them. It tells the robot how to move and respond to the world. Sometimes code is called a "program" or "software."

Coding
Writing the code for your robot. Sometimes coding is called "programming" or "software development."

Drone
Another word for "UAV."

Kinetic
Moving, not just standing still. A kinetic sculpture has parts that move.

Maker
You! Everybody and anybody who loves to make things!

Microcontroller
The "brains" that robots use. A small, programmable circuit board that designers and engineers use to make their projects interactive and autonomous. A microcontroller can have code to tell it what to do, read data from sensors that tell it what's happening in the world around it. It can connect to lights, motors, and other things that can be turned on and off. One popular kind of microcontroller is the Arduino.

MIDI
A way for electronic musical instruments (like synthesizers) to communicate with computers. MIDI stands for Musical Instrument Digital Interface. MIDI instruments produce sound by interpreting MIDI messages like "note ON," "note OFF," "pitch bend," and many more.

Pneumatic
Operated or powered by air.

Programming
See "code."

R/C
Remote-controlled.

ROV
Remotely operated vehicle.

Sensor
A piece of electronics that can see or measure something in the world and turn it into data. Sensors can sense bumps, tilt, magnetic forces, infrared signals, temperature, light, chemicals, and more!

Servo
A special type of motor that has a sensor to tell it to go to a specific place, rather than spinning round and round.

Solder
To use melted metal to join together two metal pieces.

Solenoid
A special type of a magnet that can pull a rod when you run electricity through it.

UAV
Short for "Unmanned Aerial Vehicle," it's something that flies without a human pilot on it. UAVs are also called drones, quadcopters, hexacopters, octocopters, etc.

Free for Makers like YOU!

We'll send this Learn-to-Solder pin featuring Russell the Electric Giraffe to the first 100 readers of this book who share the projects they make here:

makezine.com/lotsobots

What did you **Make**?

Whatever you make, bot or not, don't forget to tell others about it!

Tell us about your project at makezine.com/lotsobots. We want to show off what you do at Maker Faire and in our magazines. What you make will inspire other kids to make too! That's what it means to be a maker. Now that you're making, you're part of the Maker movement!

Tell us your ideas for projects. We want to hear about things you've made that will inspire others to make. (Don't tell us about things you are just thinking about making. Make it first and then tell us!)

When you tell us about your project, you'll want to explain the story behind it. Where did you get your idea, and how did it change?

Write about your project as if you're telling a smart friend how you did something. Imagine the questions they'd ask about your project. Tell them everything they need to know to make what you made. Draw pictures or take photos of each step along the way.

Save your odd bits of packaging and give it a secret superpowerful life. Turn it into a giant robot! When he was 11, Tywen Kelly made this **Styrobot** with his dad, Kevin.

You Are a Maker! Join Us!

There's a whole world of making to explore, and so many ways to be a part of the Maker movement!

Start a Maker Club.

Makers are everywhere. You may be surprised how easy it is to find them in your own school! **youngmakers.org**

Go to Maker Faire.

Events happen year-round and worldwide! Host your own Mini Maker Faire someday! **makerfaire.com**

Join Maker Camp.

Spend 6 weeks of summer making cool stuff and taking epic field trips. It's online and free! **makercamp.com**

Read Make:

Each issue has dozens of projects like the ones you saw in this book.
makezine.com

Explore Maker Shed.

The coolest, nerdiest bookstore, arts & crafts shop, electronics store, and more — all in one. We've got kits, sets, tools, and supplies.
makershed.com

Stay connected.

Makers like you share great project ideas, features, reviews, and more every day at makezine.com.

And look for extras and more in-depth material at the site for this book!
makezine.com/lotsobots

Whatever you make, never be afraid to mess up along the way, even to fail now and then. That's part of the fun you can share. Happy making!